THE
Archive Photographs
SERIES

FAIREY
AVIATION

Sir Richard Fairey at the helm of one of his 12 metre class yachts.

THE
Archive Photographs
SERIES

FAIREY
AVIATION

Compiled by
John W.R. Taylor OBE

CHALFORD

First published 1997
Copyright © John W.R. Taylor OBE, 1997

The Chalford Publishing Company
St Mary's Mill, Chalford,
Stroud, Gloucestershire, GL6 8NX

ISBN 0 7524 0684 1

Typesetting and origination by
The Chalford Publishing Company
Printed in Great Britain by
Redwood Books, Trowbridge

Front cover illustration
Refuelling at Stranraer, this Fairey IIID was the only seaplane ever
to compete in a King's Cup air race. It gained second place in 1924.

Acknowledgements

The affection in which the former Fairey Aviation Company and its products are held by so many people is a far better memorial than any book. Photographs, facts, figures and reminiscences flooded in from long-serving employees, pilots, groundcrew, historians and old friends as soon as word spread that this collection of *Archive Photographs* was to be published. The majority of the photographs came from two sources: John Fairey, son of the company's founder, who kindly made available albums of prints mounted with great care by his family and by Wilfred Broadbent, C.J. Morley and C.B. Baker, and the Fairey Group plc of today, via the author's good friend and former Fairey Aviation executive, Derek Thurgood. Special thanks go to them and to Mrs Mary Varvill, who lent the author a precious album of photographs, many taken by her father, F.G.T. Dawson, a Director of Fairey Aviation from its beginning in 1915; to Maurice Allward, J.M. Bruce, Derek James, Kenneth Munson, Norman Parker, the Revd J.D.R. Rawlings, John Stroud, Michael Stroud, Gordon Swanborough, Peter Twiss, *Flight International*, the Fleet Air Arm Museum, and the Imperial War Museum, without whose contributions there would have been worrying gaps in the record of a company that achieved so much in just forty-five years.

John W.R. Taylor,
Surbiton,
March 1997

Contents

Swordfish torpedo-bombers of the Fleet Air Arm in the still peaceful skies of 1938. By the outbreak of the Second World War, 12 of the 13 first-line Swordfish squadrons were operational on the Royal Navy's aircraft carriers HMS *Ark Royal, Courageous, Eagle, Furious* and *Glorious.* The bravery and achievements of Swordfish aircrew during the war made it the best remembered of all Fairey aircraft.

Introduction

At the age of fifteen, Charles Richard Fairey left school and became a pupil apprentice at the Jandus Electric Company of Holloway. For five evenings each week he studied at the City and Guilds (Finsbury Technical) College to qualify as an electrical engineer. Within months of completing his training, he became assistant to the manager and analytical chemist at the Finchley Power Station, still finding the time and energy to lecture at Finchley Technical College and Tottenham Polytechnic in the evenings.

Like A.V. Roe, Sydney Camm and other young men who would one day become great leaders of Britain's aviation industry, Fairey also designed and built model aeroplanes. They were so successful and won so many prizes that Gamages of Holborn, then one of London's most famous stores, paid him about £300 for rights to sell it, plus a royalty on every one bought. What he did not realise was that his model infringed patents filed by J.W. Dunne, a British pioneer whose unique sweptwing tailless aeroplanes were so stable in flight that he had created the Blair Atholl Syndicate to continue their development at Eastchurch in Kent. Fortunately, he was so impressed with the young model-maker that he waived any suggestion of licence fees, asking only that the instruction book for the model should state that it was 'licensed under J.W. Dunne's patents'. Fairey was invited to leave his work at Finchley Power Station and join Dunne at Eastchurch, which he did in 1911.

As the Royal Aero Club's official flying ground, Eastchurch attracted an exciting mixture of engineers and enthusiasts. Among the latter were three former Cambridge undergraduates who were to play important roles in C.R. Fairey's future. F.G.T. Dawson, Vincent Nicholl and Maurice Wright had built a glider and became friends of the young engineer. The association ended briefly when the First World War began and the three men volunteered as pilots for the Royal Naval Air Service. They were sent to the former Eastbourne Aviation Company's flying school, by then RNAS Eastbourne, for training, and received their Aviator's Certificates on the same day, 8 October 1914. Nicholl was given No. 936, Dawson No. 937 and Wright No. 938. All then went to war.

Meanwhile, back at Eastchurch in 1912, Fairey had joined Horace, Eustace and Oswald Short, whose company, Short Brothers, built some of Dunne's aircraft. He progressed rapidly from chief stressman to works manager and chief engineer. In the early months of the war he agreed to withdraw his application to join the Royal Flying Corps if he was given a contract to build aeroplanes in a factory of his own. Shorts needed to find companies capable of building their designs in the increasing numbers required by the Royal Naval Air Service. They offered a contract for twelve of their Model 827 seaplanes to the newly formed Fairey Aviation Company, which opened its first small drawing office at 175 Piccadilly, London.

C.R. Fairey was just twenty-eight years old. His company had a nominal

C.R. Fairey with his twin-propeller tail-first model aircraft, in 1910. After winning The Kite and Model Aeroplane Association's Challenge Cup and gold medals for 'steering', stability and distance flown, in June of that year, he demonstrated it in Hyde Park to representatives of Gamage's stores, who bought exclusive sales rights.

capital of £35,500, but the working capital was nearer £15,000. Much of it had been provided by F.G.T. Dawson, who had been invalided out of the RNAS after service in the Dardanelles to become a Fairey Aviation Director. With agreed delivery dates for the seaplanes too near for comfort, and the slender resources draining away, Fairey was able to lease part of the works occupied by the Army Motor Lorries Company in Clayton Road, Hayes, Middlesex. Some of AMLC's skilled employees, including war refugees from Belgium, joined his workforce. The seaplanes were assembled in a wooden shed in a nearby field and a water base for testing them was provided by the Admiralty at Hamble Spit on Southampton Water. They were flown by Sydney Pickles, an Australian who ranked among the ten best pilots in Britain before the war.

By mid-1916 the Short 827s had all been delivered, and work started on building 100 Sopwith 1½-Strutter two-seat fighters for the RNAS. The drawing office was busy designing a large twin-engined fighter known as the F.2, which became the first Fairey type to fly. Its survivability in combat against the agile single-seat fighters of the time was questionable, so it was abandoned, and Fairey decided to concentrate on the seaplanes for which he knew there was a demand.

A major problem with all seaplane operations was that the aircraft had to be lowered over the side of ships carrying them, onto the water, before they could take off. They were often damaged, especially if the water was not calm. In an effort to find a solution, the Admiralty fitted a wooden flying deck on the former Cunard passenger liner *Campania*. The Fairey Company designed a two-seat patrol seaplane, named Campania after the ship, that could take off directly from the 200 ft deck by means of a trolley that fitted under the floats. The aircraft carrier had been born, although the seaplanes had to land on the water for recovery.

Admiralty orders for at least 100 Campanias represented the first production contracts for an all-Fairey design. At the same time, the company redesigned the single-seat Sopwith Baby seaplane into the Hamble Baby, for anti-submarine patrol and light bombing. What made it specially significant was that it introduced the Fairey patent camber-changing gear, a form of wing flaps that was to enhance the performance of Fairey designs for two decades.

As its wartime commitments grew, Fairey Aviation gained control of the entire Clayton Road factory and added a drawing office and experimental workshops at North Hyde Road, Hayes, beside its assembly shed. However, it still needed improved facilities. With a Government loan of £20,000, a purpose-built plant was erected at North Hyde Road, including an assembly shop with a span of 90 ft and height of 24 ft. Opened in the spring of 1918, it expanded gradually into the company's life-long administrative headquarters and main factory.

Several experimental aircraft were designed and built before the Fairey IIIA light bomber landplane entered production. The war ended before it could become operational, but some related Fairey IIICs, on floats, were taken to Archangel on the carrier HMS *Pegasus* as a unit of the North Russian Expeditionary Force in 1919, and were probably the best seaplanes designed anywhere during 1914-18.

Production contracts for military aircraft were cancelled overnight when the First World War ended and several famous British manufacturers went out of business. To keep the Hayes factory at work, a company named Fairey and

Eastchurch, 1911. Fairey stands by the wingroot of J.W. Dunne's D.8 tailless biplane, while working for the Blair Atholl Syndicate. His models had infringed Dunne's patented method of attaining stability by varying the camber of the wings from root to tip. Instead of demanding payment, Dunne had invited Fairey to join him.

Probationary Flight Sub-Lieutenant F.G.T. ('Wuffy') Dawson in Short S.62. He had joined the RNAS with Vincent Nicholl and Maurice Wright at the start of the First World War. All three later became Fairey Aviation Directors. Dawson helped to finance the new company; Nicholl also served as test pilot from the end of the war until 1923.

Charles Ltd was formed to build motor car bodies. In the event it was not needed. By the time Fairey Aviation had been put into voluntary liquidation and re-formed in 1921, it had flown the prototype IIID seaplane. This proved so successful that a first batch of fifty was ordered for the Royal Air Force, which had come into existence in April 1918 by the amalgamation of the RFC and the RNAS.

Fairey was also awarded the RAF's first peacetime contracts for a new fighter, the sea-going Flycatcher, and light day bomber, the Fawn. Recalling the latter in a lecture in 1931, C.R. Fairey commented that: 'The net result of the drawing up of this aircraft specification, and of ineffective co-operation with the designer, resulted in an inferior machine after five years and much expenditure of money'. Typically, all the Fawn's fuel had to be carried in two massive tanks above the wing centre-section, to satisfy the Air Ministry's fire-protection regulations. Little wonder that its maximum speed of 114 mph was less than that of a D.H.9A of the First World War.

One of the favourite stories of inter-war British aviation tells how R.J. Mitchell of Supermarine watched America's Curtiss racing seaplane outfly his old-fashioned Sea Lion flying-boat in the 1923 race for the coveted Schneider Trophy, got approval to design his elegant S.5, S.6 and S.6B monoplanes that eventually won the trophy outright for Britain, and evolved from them the Spitfire fighter. This is true, but there is a parallel story of equal significance.

Impressed, like Mitchell, by the winning Curtiss R-3, C.R. Fairey offered to design for the Air Ministry a bomber embodying the same beautifully streamlined form, made possible by a 450 hp Curtiss D-12 engine of minimal frontal area. Bureaucracy was not interested, so Fairey scraped together as much of his own and the company's money as possible, went to the USA, and came back with a D-12 plus licence rights for the engine, its Curtiss-Reed metal propeller, high-efficiency wing sections and wing-surface radiators. His chief designer, the Belgian Marcel Lobelle, and P.A. Ralli of the technical department, used this American expertise to produce the prototype Fox two-seat day bomber.

First flown by Norman Macmillan on 3 January 1925, it proved 50 mph faster than a Fawn, carrying the same bomb load. Official resistance to its foreign engine and internal fuel tanks was not overcome until Sir Hugh Trenchard, the respected 'father of the RAF', saw it demonstrated, was told by Macmillan that it was one of the easiest and most viceless aeroplanes he had flown, and ordered enough Foxes for one full squadron. During subsequent defence exercises they were not allowed to operate to their full capability or the RAF's fighters of the time could not have caught them.

After a while the Curtiss engines were replaced by Rolls-Royce Kestrels, designed to fit into similar close cowlings; but no more RAF Foxes were ordered. The same official coolness was shown to the Firefly, a single-seat fighter counterpart of the Fox, first flown with a D-12 engine. Both types were adopted enthusiastically by the Belgian Air Force, in improved versions with Kestrel engines, leading to the formation of an associated company named Avions Fairey at Gosselies in Belgium, to assemble them. The Curtiss engines fitted to the first Firefly and early Foxes were imported from the USA and,

The workshops and slipways at Hamble Spit, from which Fairey-built Short 827 seaplanes were tested in 1916, had to be built on concrete stilts as the marshy ground was subject to flooding.

Four Short 827 seaplanes undergoing final assembly at Hamble in 1916. The wooden buildings leased to Fairey Aviation by the Admiralty were surprisingly roomy.

although the name Fairey Felix was applied to them, planned manufacture in Britain never took place. C.R. Fairey avoided becoming too disenchanted with his treatment by purchasing large and beautiful racing yachts, of which he became one of Britain's outstanding helmsmen, while remaining firmly at the helm of his company in Hayes.

The Fairey III series reached its zenith with the IIIF. First flown in 1926, it went into production as a two-seat, general-purpose aircraft for the RAF and as a three-seat, spotter-reconnaissance aircraft for its sea-going squadrons, known since April 1924 as the Fleet Air Arm. By 1932 a total of 622 had been completed, plus more than 250 Seal and Gordon variants. Except for the Hawker Hart and its derivatives, no other British military aircraft was delivered in such numbers between the end of the First World War and the start of Britain's rearmament programme in the mid-1930s.

The Fairey Aviation Company Ltd was registered as a public company, with a nominal capital of £500,000, on 5 March 1929. Its assets at Hayes and Hamble were valued at £615,486. To them was added in the following year a new flight test centre at Harmondsworth, a few miles from Hayes, on a piece of land that C.R. Fairey had bought for £15,000. Known later as the Great West Aerodrome, it replaced Northolt and remained in company use until 1944 when it was absorbed into what was to become the world's No. 1 international civil airport of Heathrow. Fairey Aviation then removed its flight testing to nearby Heston and to White Waltham in Berkshire.

The period spanned by the years at the Great West Aerodrome was dramatic. It began well. In the 1920s, Fairey IIIDs and IIIFs had shown the flag and developed the RAF's experience of long distance flying with a succession of formation flights from Cairo to Cape Town and back, and over other long pioneering routes. Fairey received Air Ministry contracts to construct what were described as two 'Postal Aircraft, to study ways of increasing range by all practical means'. In reality, the aim was record-breaking and, in 1933, the second of these Long-range Monoplanes exceeded the previous world's absolute distance record by 298 miles. For a brief period, Britain held all three major records for speed, height and distance.

Fairey went on to design the RAF's first cantilever monoplane heavy bomber, the Hendon; but only fourteen were built in the big Willys-Overland Crossley motor factory at Heaton Chapel, Stockport, that Fairey took over in 1935 to meet the demands of rearmament for a war that seemed inevitable. All but the earliest aircraft built in this factory were flown from Manchester's Ringway Airport.

The two types of aircraft ordered into large-scale production in the mid-1930s provoked heated controversy during the Second World War. The Battle bomber, produced at Heaton Chapel and in the Austin Motors' Shadow Factory, was described by the editor of *The Aeroplane* magazine in 1936 as an aircraft that would show the Americans how to build a bomber. Sadly, it was outdated and almost defenceless against attack by Luftwaffe Messerschmitt Bf 109s by the time its crews, with extreme bravery, attempted to stem the German advance through Belgium and France in 1940.

The Swordfish torpedo-spotter-reconnaissance biplane of the Fleet Air Arm, built initially at Hayes and then by Blackburn Aircraft, seemed an anachronism from the moment it appeared in 1934. Its planned replacement, the Albacore, had already flown before the outbreak of war, but the Swordfish outlived it in

The new factory built beside North Hyde Road, Hayes, in 1918. The works entrance is in the foreground, with the main offices beyond the chimney. To the south of the site stood the original wooden assembly building, on the edge of a primitive grass 'aerodrome'. Testing of landplanes was normally done at Northolt.

First flown on 3 January 1925, the prototype Fox was the aircraft with which Fairey Aviation revolutionised British ideas of military aircraft design. The slim, clean lines made possible by its American Curtiss D-12 engine increased the speed of RAF day bombers by 50 mph, but only 28 D-12 Foxes were built, for use by a single squadron of the Royal Air Force.

service. It sank more enemy shipping than any other Allied aircraft, put out of action many of the finest ships of the Italian fleet at Taranto in 1940, damaged the mighty German battleship *Bismarck* so effectively that it could be caught and sunk by the Royal Navy in 1941, and much more.

As the war progressed, Albacores and Firefly monoplane fighters were produced at Hayes, Fulmar fighters and Barracuda torpedo bombers at Heaton Chapel. Errwood Park Shadow Factory, adjacent to Heaton Chapel and managed by Fairey, built Bristol Beaufighters and Handley Page Halifax bombers under subcontract. The newly knighted Sir Richard Fairey could only watch all this from a distance. For most of the war years he was a key member of the British Air Commission in Washington, latterly as Director General. Its work in co-ordinating design and production activities between the UK and USA, and ensuring that the best available aircraft were ordered for the Royal Air Force and Fleet Air Arm (which had reverted to Admiralty control in 1937) was vital to the war effort, and gained him the US Medal of Freedom, but imposed lasting strain on his health.

This time, production contracts were not all cancelled when the war ended. The Barracuda continued to be manufactured until October 1947 and the Firefly until March 1956, but their successor, the anti-submarine Gannet, was to prove the last Fairey Aviation production aircraft.

Paradoxically, the years during which it was on the assembly lines seemed to promise the company a future at the forefront of new technology. Fairey became an early leader in British missile development and used its growing capability to design the remarkable little Delta One as a research vehicle for a fighter that might be launched vertically from ships of the Royal Navy. It was followed by two entirely different and advanced, supersonic Delta Twos. Flown by test pilot Peter Twiss, one of them set a world's absolute speed record of 1,132 mph (Mach

1.72) on 10 March 1956, beating the previous record by an unprecedented 38 per cent.

It had been hoped that expertise gained with the Delta Two might lead to work on a supersonic fighter, but the Government had ordered the English Electric Lightning to meet that requirement and was already planning a short sighted defence policy that envisaged the future replacement of manned fighters and bombers with missiles. This was a major disappointment, but Fairey had also established a reputation at the leading edge of progress in the new helicopter industry from 1946. Led initially by Dr J.A.J. Bennett, it pioneered a type of aircraft that took off as a helicopter and converted in flight to cruise as an autogyro. The initial four-seat Gyrodyne set a world's helicopter speed record of 124.3 mph in June 1948, before the design progressed to the Jet Gyrodyne with a pressure-jet unit at each rotor blade tip.

Soon, White Waltham was home to the diminutive two-seat multi-role Ultra-light and large twin-turboprop Rotodyne that attracted interest from British European Airways and New York Airways as a potential 54/65-seat commuter airbus. Both utilised tip-jets, which presented noise pollution problems, but a far more crucial threat came from the Government. Under its policy of 'rationalising' the aircraft industry, Fairey helicopter activities were taken over by Westland on 8 February 1960. Earlier, Fairey Aviation Company had become Fairey Company Ltd, a holding company for a variety of subsidiaries. One of them was Fairey Aviation Ltd, which was also absorbed by Westland in 1960.

Sir Richard Fairey believed that 'If the world does not please you, you can change it'. He died too soon to exert any influence on Government policy that has slowly destroyed Britain's leadership in the air.

Wings folded in salute, Gannets of No. 703X Flight bid farewell to guests who attended the ceremony of handing over their log books at Royal Naval Air Station, Ford, Sussex, on 5 April 1954. This unit was responsible for intensive flight trials of what was to be the last type of combat aircraft built by Fairey Aviation.

One
War and Peace

From the start, the Fairey Company devoted most of its effort to producing seaplanes and aircraft for service from ships and land bases of the Royal Navy. It also thought big. Together with the Gramophone Company (later HMV), its neighbour in Clayton Road, it made parts for the world's largest aeroplane, the 142 ft span Kennedy Giant, in 1916-17.

The Giant's designer, Chessborough J.H. Mackenzie-Kennedy, had been associated with Igor Sikorsky in the construction of the first four-engined aeroplane, in Russia, before the war. He persuaded the British War Office that the Royal Flying Corps might make good use of a bomber like Sikorsky's highly successful Ilya Mourometz.

Because of its size, the aircraft had to be assembled in the open at Northolt aerodrome. With four 200 hp Salmson engines, in tandem pairs, it was clearly underpowered, but the highly-skilled Frank Courtney attempted to persuade it into the air in late 1917. At full throttle, the Giant rumbled down a gentle slope and the four mainwheels briefly left the grass. The tailskid refused to follow. When the slope ended, the wheels resumed their rumble on the ground and the aircraft came to a stop, still at full throttle. It underwent many changes during the following years and presented a fascinating sight to pilots flying in and out of Northolt, but never again attempted to fly - and gradually crumbled away.

Previous pages: The Kennedy Giant, in its initial form, at Northolt in 1917.

The Fairey F.2, seen here with its wings folded, was the first aeroplane both designed and built by Fairey Aviation. After brief hops on the 'aerodrome' by the assembly shed at Hayes, it was taken to Northolt and flown by Sydney Pickles on 17 May 1917. Two 190 hp Rolls-Royce Falcon engines gave it a speed of 93 mph. The crew of three included gunners with Lewis machine guns in the nose and behind the wings.

The Admiralty specification to which the F.2 was designed was clearly unimaginative. No large, relatively slow fighter would survive in combat with Germany's agile single-seaters of 1917. The F.2's potential as an anti-Zeppelin night fighter or strategic bomber seems to have been overlooked.

The patriotically-decorated drawing office at Hayes in 1919. The chief draughtsman was A.C. Barlow, formerly of Short Brothers.

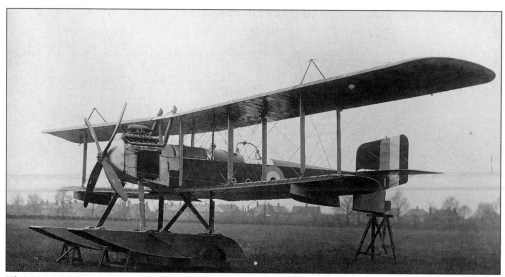

The prototype Campania had a 250 hp Rolls-Royce Mk IV (Eagle IV) engine and was designed specifically for operation from carriers of the Royal Navy. After testing at the Marine Experimental Aircraft Depot on the Isle of Grain, near Sheerness in Kent, it made a remarkable non-stop flight to Scapa Flow, piloted by Sqn Cdr Maurice Wright, who later became a Director of Fairey Aviation.

A Campania takes off from the Royal Navy carrier after which it was named. The four-wheel trolley under its floats could be stopped at the end of the deck and retrieved for further use. Twenty-five of the 62 Campanias completed had 260 hp Sunbeam Maori engines; the rest had Eagles of up to 345 hp. They operated from HMS *Nairana* and *Pegasus*, as well as from *Campania* and shore stations.

The pilot's cockpit of N1004, the fifth Campania built, photographed during the aircraft's flight testing at the Isle of Grain on 29 September 1917. Behind the control wheel is an altimeter, with an oil pressure gauge, airspeed indicator and rev counter to the right and a compass above the coaming.

The Hamble Baby was based on the Sopwith Baby but had new wings, of which the entire trailing-edge was hinged. Lowered for take-off, this Fairey patent camber-changing gear, a pioneer form of wing flaps, increased lift to such a degree that the seaplane lifted off the water quickly even when carrying two 65 lb bombs. The hinged surfaces also worked differentially as ailerons. Hamble Babies were used primarily for anti-submarine patrol and occasional bombing raids.

The N9 two-seat seaplane was designed and built in 1917 for potential operation from seaplane carriers. It failed to win a contract, but was strengthened for pioneering trials of a launch catapult on HMS *Slinger*. Its engine was a 200 hp Rolls-Royce Falcon I.

N9 ready for launch on the catapult rails of HMS *Slinger* at the Isle of Grain experimental station in June 1918. The pilot for most of the trials was Lt Col H.R. Busteed.

The trials of the Armstrong Whitworth compressed-air catapult were successful, although the launch speed of around 40 mph was only 2 mph above the N9's stalling speed. It was to be another 7 years before the first in-service aircraft, a Fairey IIID, was catapulted from a warship, the cruiser HMS *Vindictive*.

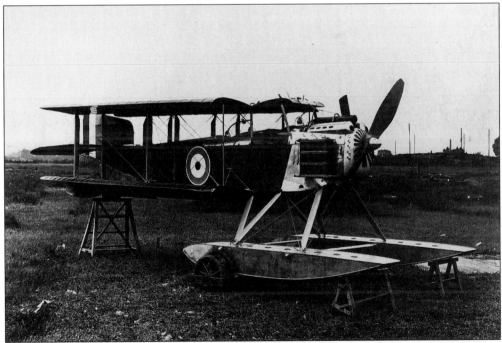

The second experimental seaplane built by Fairey to Admiralty specification N.2(a), in 1917, was the N10, with a 260 hp Maori II engine. Known later as the Fairey III, it was the starting point for the succession of widely used Fairey Series III aircraft.

The machine shop at the Clayton Road factory in 1917. The production manager was young Wilfred Broadbent, who was also in charge of the company's experimental workshop. He became a Director of the company from 1929 until his retirement in 1958.

Late in 1917, the floats of the N10 seaplane were exchanged for wheels and it was redesignated Fairey IIIA. The Royal Naval Air Service ordered fifty to replace its old Sopwith $1\frac{1}{2}$-Strutters as shipborne two-seat bombers. Fourteen of them (the second production IIIA is illustrated) were fitted with wheels; the rest had skids, which were believed to ensure a straighter run during take-off on deck.

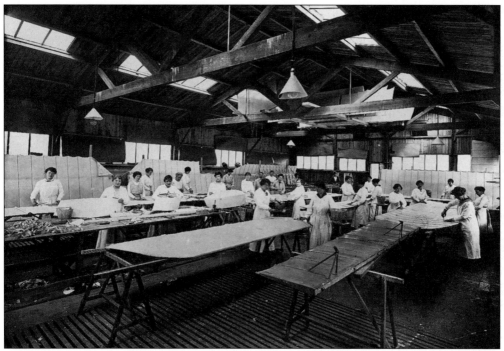

As in most companies, wings were covered with fabric and doped by women workers. This was the wing shop at Clayton Road in 1917.

View from Clayton Road of Fairey Aviation's new works at Hayes, opened in 1918 beside North Hyde Road. In the foreground is the Great Western Railway's main line from London to Reading and Oxford. The factory expanded greatly as time passed and remained the company's headquarters throughout its life.

The IIIB was a seaplane bomber variant of the IIIA with the span of the upper wings extended from 46 ft 2 in to 62 ft 9 in. Like the IIIA it carried bombs on underfuselage racks. Fewer than 30 were built, of which some flew mine-spotting patrols from the seaplane station at Westgate-on-Sea, Kent, before the war ended.

In 1918-19, most uncompleted Fairey IIIBs in the erecting shop at Hayes were converted into IIICs, capable of both bombing and reconnaissance duties. Only 36 were delivered, but their 375 hp Rolls-Royce Eagle VIII engines helped to make them the best seaplanes of the First World War, paving the way for the outstanding IIID in 1920.

Fairey Aviation bought back the N10 from the Admiralty and entered it for the first post-war contest for the Schneider Trophy at Bournemouth, in September 1919. The wings were cropped from 46 ft to 28 ft span, the Maori engine replaced with a 450 hp Napier Lion and the front cockpit covered. Poor visibility on the day of the contest caused such havoc that it was declared void.

This Fairey IIIC was fitted with three-bay wings, a transparent canopy over its cockpits and a Rolls-Royce Eagle VIII engine, as Fairey's entry for the 1919 competition to win a £10,000 *Daily Mail* prize for the first non-stop Atlantic flight. After Alcock and Brown's successful flight in a Vickers Vimy, the IIIC was shipped across and became G-CYCF on the Canadian Register.

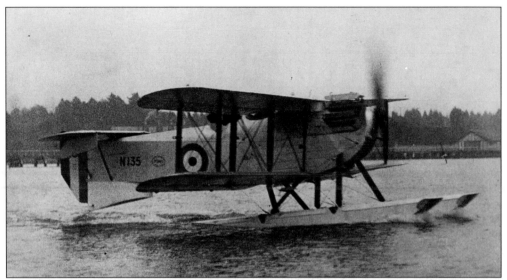

The Pintail was designed to meet RAF Specification XXI for a two-seat fighter-reconnaissance amphibian able to operate from land, water or carrier deck. Its unique tail surfaces gave the gunner a clear field of fire rearward. N135 was the last of three prototypes with different forms of undercarriage. Its wheels were fixed within the floats, a configuration that was adopted for three Pintails delivered to Japan. These had the upper wing raised by 9 in to improve the pilot's view.

A Transatlantic Load Carrier version of the Fairey IIID, with extra fuel tanks in the floats, was built for the Portuguese Government to attempt the first flight from Portugal to Brazil, in stages, over the South Atlantic. It left Lisbon on 30 March 1922, flown by Cdr Sacadura Cabral and Capt Gago Coutinho. On 18 April they navigated precisely over the $11\frac{1}{2}$ hour leg to the tiny 22 acre St Paul's Rocks, where a cruiser waited to refuel them. Rough seas broke off a float as they landed and the seaplane sank. They were sent a replacement, but its engine cut and they were nearly lost after ditching in the sea. The two men eventually reached Rio de Janeiro on 17 June in this third IIID, having flown about 5,030 miles.

Pilots undertook long distance flights with increasing confidence in the early 1920s. On 19 May 1924, Wing Cdr S.J. Goble, DSO OBE DSC, and Fl Off I.E. McIntyre, RAAF, returned to base after flying around the coastline of Australia in this Fairey IIID: 8,568 miles in 90 hours.

Fairy IIIDs on the foredeck of HM Seaplane Carrier *Ark Royal*. In the foreground are two cranes used to lower them into the water and retrieve them after flight.

This Fairey IIID took second place in the 1924 King's Cup air race, covering the 950 mile course at 108 mph. Refuelling time on the water at Stranraer was a mere 12 minutes. The pilot was Norman Macmillan, who had replaced Vincent Nicholl as chief test pilot. His engineer passenger was E.O. Tips, later the designer of Tipsy light aircraft and Manager of Avions Fairey in Belgium.

Four IIIDs of the Cape Flight completed the Royal Air Force's first long distance formation flight in 1926. Leaving Cairo on 1 March, they flew 11,000 miles without major problems, to Cape Town and back. At Abu Qir they exchanged their wheels for floats before returning home to Lee-on-Solent in England on 21 June. The Flight was led by Wing Cdr C.W.H. Pulford, OBE AFC.

Four Fairey IIID seaplanes were sold to the Royal Netherlands Naval Air Service in 1924, for operation in the Dutch East Indies. They are shown at Hamble, awaiting delivery.

In the last year of war, the Admiralty had ordered from Fairey two fleet co-operation and reconnaissance flying boats that were to be the largest in the world. Wing span was to be 139 ft, weight more than 30,000 lb including 1,000 lb of bombs and six gun positions, and patrol endurance at least seven hours. Because of their size, manufacture was subcontracted to Dick Kerr and Company at Lytham St Annes, Lancashire, with Linton Hope-Fairey hulls made by experienced boat builders. Post-war problems delayed the assembly and first flight of N119 *Atalanta* (illustrated) until 4 July 1923, powered by four 650 hp Rolls-Royce Condor engines in tandem pairs. Her sister, N129 *Titania*, flew on 24 July 1925; but the RAF could find no use for such large aircraft by then.

Seventy production Fairey Fawns were the first light bombers of post-war design built for the Royal Air Force. To conform with Air Ministry regulations, their fuel was carried in two large tanks above the wing centre-section, seen clearly in this view of a partially completed fuselage. They restricted its maximum speed to 114 mph.

Fawn IIs, with 470 hp Napier Lion engines, of No. 12 Squadron, Royal Air Force, rehearsing for the Wing Drill by a formation of 36 day bombers that was a feature of the 1925 and 1926 Hendon Air Displays.

Teams from Fairey Aviation always played a major role in local sporting events, including football, rugby and ladies' hockey. The winners of the Hayes and District Cricket League in 1920 and 1921 are seen here with C.R. Fairey (behind the cup) and Wilfred Broadbent (second left, centre row).

Sir Samuel Hoare, Secretary of State for Air, and the Rt Hon Winston Churchill viewing a Fairey Flycatcher single-seat fighter of the Fleet Air Arm at the 1923 SBAC Show.

Three Flycatchers line up for take-off from HMS *Furious*, ahead of three Fairey IIIFs in 1930. A cocked-up rear fuselage did not make the Flycatcher the most handsome of aircraft, but its 400 hp Armstrong Siddeley Jaguar engine gave a top speed of 133 mph; it had unrivalled agility and its Fairey camber-changing gear (full-span flaps, dating back to the Hamble Baby) provided great deck-landing performance.

An unusual view of a Flycatcher on the deck lift of HMS *Furious*, with another parked on the hangar deck below and beyond it.

Flycatchers of the 'slip-flights' on HMS *Furious*, *Courageous* and *Glorious* took off straight from their hangar on a 60 ft runway built over the bows of the ships. The Wing Cdr's bridge can be seen on the port side of the main flight deck, with the Captain's bridge on the starboard side.

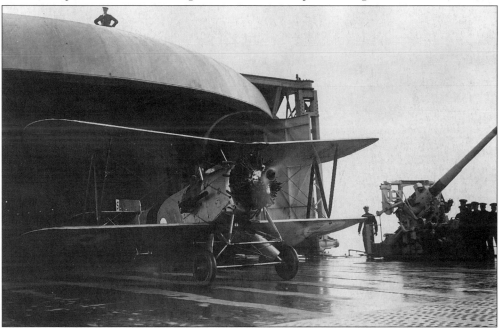

A 'slip-flight' Flycatcher taking off from HMS *Furious*. One of its two synchronised Vickers machine guns can be seen on the side of the fuselage; four 20 lb bombs could be carried under the wings.

Even Flycatchers did not always deck land correctly, as Lt Hunt discovered when he headed for the palisades at the side of the deck on 11 February 1929.

A Flycatcher seaplane over the Fleet in Gibraltar harbour. The large floats reduced the fighter's speed by only 7 mph.

Immaculate formation flying has always been a hallmark of Naval units, as these Flycatchers of No. 403 Flight demonstrated when they flew on China Station as members of the Catapult Flight, 5th Cruiser Squadron.

The Flycatcher was the first Fleet Air Arm fighter stressed for catapult launch from cruisers and other warships. Construction was of wood with fabric covering, except for the steel-tube forward and centre fuselage structure. With a wing span of only 29 ft, it did not need folding wings to use carrier deck lifts but could be dismantled, with no section exceeding 13 ft 6 in, for stowage on board ship.

More like the later Firefly I than the original Flycatcher, the Flycatcher II was first flown by Norman Macmillan on 4 October 1926 with a Jaguar engine and wooden wings. In August 1927 it was tested on floats and, by May 1928, had a 480 hp Bristol Mercury engine, metal wings and speed of 153 mph as a landplane. However, it remained a one-off prototype.

Nominally a long range reconnaissance aircraft, to Air Ministry Specification 44/22, the Fremantle seaplane was, in fact, built to attempt the first round-the-world flight. It was large, with a wing span of 69 ft 2 in, and a mahogany-planked cabin roomy enough for the crew to stand upright, sleep, eat and navigate in comfort. The pilot's cockpit was still open. Fuel for the 650 hp Condor engine was carried in a tank above the top wing and in the main floats.

Too late to challenge the US Douglas World Cruisers that flew round the world in 1924, the Fremantle was used for radio navigation development by the Royal Aircraft Establishment.

The three Ferret prototypes marked a major milestone in Fairey manufacture, as they had all-metal structure, still with fabric covering. The Mk. II, illustrated, was intended as a three-seat reconnaissance aircraft for the Fleet Air Arm, powered by a 425 hp Bristol Jupiter engine.

Modesty, one of the beautiful 12 metre yachts raced with considerable success by C.R. Fairey during the inter-war years.

Two

Faster and Farther

The Fairey IIIC and IIID had proved outstanding aircraft in their time. The Fawn and Flycatcher reflected the company's growing reputation by capturing the first RAF post-war production contracts for light bombers and fighters. C.R. Fairey, who had been honoured with the MBE in 1920, had no doubt that he could do better. In an inspired move he imported fifty Curtiss D-12 aero-engines from the USA, used some of them in his finely-streamlined two-seat Foxes, and raised the speed of RAF day bombers by more than 35 per cent. Rolls-Royce's Kestrel engine soon offered a similar small frontal area to the D-12, which it replaced in the Fox. Packed into the slim noses of a whole generation of new combat aircraft, it pointed the way to the Merlin, the most important aero-engine of the Second World War.

It is often forgotten how many Fairey aircraft flew with the Royal Air Force and Fleet Air Arm in the 1920s and '30s. A few 'specials', like the record-breaking Long-range Monoplane, were spectacular; some like the unwanted Fantôme, were truly beautiful. Most, like the unglamorous IIIFs, simply gave reliable, little-publicised service during twenty years of near-peace.

The prototype Fairey Fox revolutionised ideas on combat aircraft design in the UK. Its 450 hp American Curtiss D-12 engine and clean lines made it 50 mph faster than the Fawn, then the RAF's standard post-war day bomber. It was first flown on 3 January 1925.

Head-on, the slim lines made possible by the D-12 engine became even more apparent. Air Chief Marshal Sir Hugh Trenchard, Chief of Air Staff, wasted no time in ordering a squadron of Fox day bombers for the Royal Air Force.

FAIREY FOX

59th Anniversary of the
First Flight of the Fairey Fox –
3 January 1984

Fairey Foxes of No.12 Squadron
over Salisbury Plain during
Autumn Exercises 1927

In the series of Royal Air Force Museum covers commemorating famous RAF aircraft, the Fairey Fox was featured on the 59th anniversary of its first flight. John Young's painting shows D-12 engined Foxes of No. 12 Squadron over Salisbury Plain during Autumn Exercises in 1927.

Proud of its new equipment, No. 12 Squadron designed a unit badge depicting a fox, which has been carried on its aircraft, in various styles, to this day.

The 12 Squadron fox badge is visible on the fin of this Fox IA, re-engined with a 480 hp Rolls-Royce F.XIIA (Kestrel IIA). This view shows clearly the low-drag form of the rear cockpit, with a Fairey high speed gun mounting replacing the traditional Scarff ring.

Shaken by the performance of the Fox, the Air Ministry issued a Specification for a new day bomber in 1926, but did not invite Fairey to tender. It protested and was sent a belated copy of the Specification. By the time its much redesigned Fox IIM flew, on 25 October 1929, the Hawker Hart had already been ordered. With its all-steel structure, instead of the Mk I's wood, the IIM had immediate appeal to the Belgian Air Force. Avions Fairey built a factory at Gosselies, Belgium, in which to produce it, and eventually delivered a total of 189 in 10 progressively improved forms. Illustrated is one of 12 Fox IIMs supplied from Hayes in 1932 in advance of deliveries from Gosselies.

The Belgian-built Fox Trainer was unique in having a 340 hp Armstrong Siddeley Serval radial engine and dual controls, with a retractable hood on the rear cockpit for blind flying instruction. Only one was built and was re-engined in 1934 with a Kestrel IIMS, as the prototype Fox IIIS.

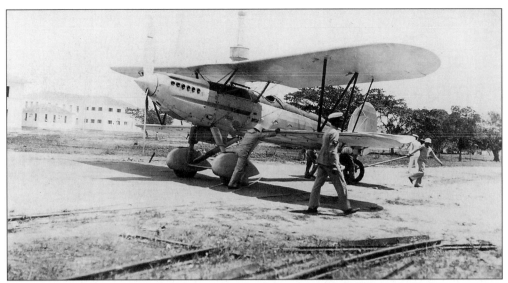

The Fox IV demonstrator G-ABYY, with a 525 hp supercharged Kestrel IIMS engine, was shipped out to the Far East in search of customers. Marketed as a reconnaissance-fighter with two forward-firing machine guns, it was flown by Col Wu (foreground) of the Kwangsi Air Force during a sales tour to Liuchow.

Use of an American engine in the original Fox I day bomber had led to a great deal of criticism of Fairey in UK aviation circles. When the company acquired licence rights in the Curtiss D-12, called it the Fairey Felix and fitted it in the prototype Firefly single-seat fighter, the Air Ministry made it clear that it would be interested in buying more than one only if the D-12 was replaced by a Rolls-Royce engine.

When Fairey had built the Firefly II with a Rolls-Royce Kestrel engine, the Air Ministry still did not order it; but Belgium did. This first batch of Hayes-built Fireflies was handed over to the Belgian Government at Harmondsworth (now Heathrow) on 6 July 1931. Following delivery of 25 aircraft from the UK, a further 63 were assembled at Gosselies.

Last of Fairey's Series IIIs, the IIIF was built as a three-seat spotter-reconnaissance aircraft for the Fleet Air Arm and two-seat general-purpose aircraft for the Royal Air Force. These IIIF Mk Is, photographed at Kano in Nigeria, formed part of a four-plane formation led by Sqn Ldr F.J. Vincent that flew from Cairo to West Africa and back in 1929. Early IIIFs had horn-balanced rudders.

A IIIF Mk IIIM of No. 822 Squadron, Fleet Air Arm, over HMS *Furious*. The suffix M indicated that it was all-metal, whereas early IIIFs had wooden wings. The curved fin and 570 hp Lion XIA engine were also standard on later versions for UK service.

RAF IIIF Mk IVMs in the Hayes erecting shop in 1929. J9160 was one of several aircraft with special passenger seating for VIP transport by No. 24 (Communications) Squadron, and was allocated for use by Sir Philip Sassoon, Under Secretary of State for Air.

The intricate all-metal structure of the Fairey IIIF is apparent in this photograph of an aircraft at Hayes before the metal engine cowling and fabric covering were applied.

Harry Masser, in the cockpit, and colleagues check the wiring and shock-absorbers of a IIIF. It was the first type designed to absorb a vertical descent rate of 12 ft/sec, a new Air Ministry requirement for deck landing.

A newly completed IIIF airframe packed for transport from Hayes to Northolt. As can be seen, the well proven Fairey camber-changing gear continued in use. Cross-deck arrester wires were not yet fitted to RN carriers, so no arrester hook was needed.

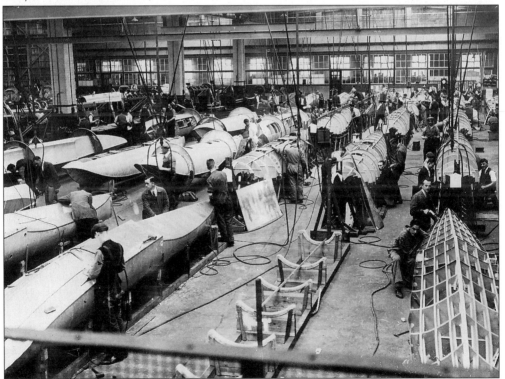

Fairey Aviation built its own floats, initially of wood, later of metal. One early IIIF was fitted with specially strengthened floats for experiments in taking off and landing unaided on the deck of HMS *Furious*. In fact, it needed rocking and pushing by deck crews before it would move forward and usually landed with a shower of sparks in anything but a straight line.

Hayes works, photographed from a de Havilland Puss Moth, on 28 December 1930. The mainline railway can be seen running under the road bridge to the north.

A total of 379 Fairey IIIFs was delivered to the Fleet Air Arm, making them the most widely used of all British Naval aircraft in the inter-war years.

These IIIF Mk IIIBs of No. 824 Squadron were based on HMS *Eagle* with the Mediterranean Fleet.

Catapulting a IIIF Mk IIIB from a Royal Navy warship. Aircraft specially strengthened for this role flew from the battleship *Valiant*, the battle cruiser *Hood* and the cruisers *Dorsetshire*, *Exeter*, *Norfolk* and *York*.

After flying to Khartoum in February 1929, this IIIF Mk IVC/M of the Royal Air Force is having its wheels exchanged for floats to permit operation from the Nile.

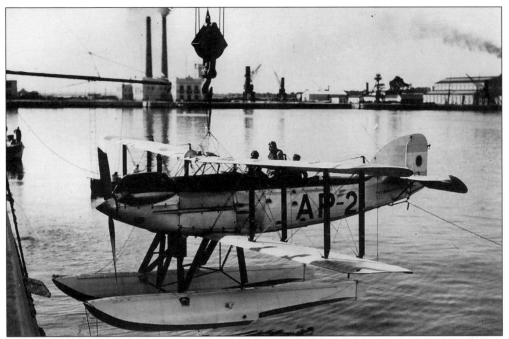

The six IIIF Mk IIIMs sold to the Argentine Government were delivered with 450 hp Lorraine Ed12 engines, with helmet cowling over the 12 broad-arrow cylinders. When possible, seaplanes being hoisted from the water were tethered by ropes from shore or boats; on other occasions the propeller was kept ticking over to permit sufficient control to prevent the aircraft swinging and being damaged.

The hazards of taking off and landing on the open sea are suggested by the spare float visible on the crowded deck space behind this IIIF Mk IIIB, stowed on the midships catapult of a Kent class cruiser.

Increased dihedral on its wings for improved stability, and a four-blade propeller, identify the Fairey Queen, a version of the IIIF intended for radio-controlled pilotless operation as a gunnery target for the Fleet. The first two launches from a catapult on HMS *Valiant*, on 30 January and 19 April 1932, ended in crashes into the sea after 18 seconds and 25 seconds of flight. Two hours after launch on the third flight, the Fairey Queen was landed on rough water with no damage from all the gunfire aimed at it by the Home Fleet. The Mediterranean Fleet did rather better in May 1933, by taking 20 minutes to shoot down the unpiloted aircraft from 8,000 ft. This dubious success led to development of the de Havilland Queen Bee target, based on the Tiger Moth.

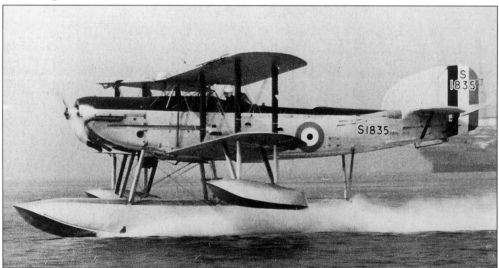

Many experimental versions of the IIIF were produced, including this IIIF Mk IIIB fitted with a long central float and two underwing stabilising floats. It also had stainless steel wings that had already been subjected to 93 hours of flying to compare their resistance to corrosion with that of standard steel wings. After a further 92 hours of flying, they spent $2\frac{1}{2}$ hours immersed in sea water after an accident to S1835. When cleaned, they were judged to be far better than any standard wings would have been after such treatment.

This photograph of the first Fairey Long-range Monoplane, J9479, in a vast empty sky seems to convey something of the loneliness of the long distance flyer. Built to set a new world distance record, its RAF crew, Sqn Ldr A.G. Jones-Williams, MC, and Flt Lt N.H. Jenkins, made the first non-stop flight from the UK to India in April 1929 but landed 336 miles short of the existing record. They were killed in a second attempt on 16 December.

The second Long-range Monoplane had 82 ft span wings, a slim fuselage and 570 hp Napier Lion XIA engine like the first. Fuel load was increased to 1,147 gallons, the wheels were enclosed in large spats and equipment was improved, including addition of an automatic pilot. Range was estimated at 5,550 miles at around 95 mph, which was adequate to beat the distance record, by then 5,011 miles.

Fairey's chief test pilot since 1930, Flt Lt Chris Staniland made the first flight in K1991, the second Long-range Monoplane, on 30 June 1931. Well known as a racing car driver at Brooklands, he did much to ensure the quality of Fairey aircraft until his death in a flying accident in June 1942.

Sqn Ldr O.R. Gayford, DFC (right) was captain on K1991's long-distance flights. His companion in this photograph, Flt Lt D.L.G. Bett, was the navigator on an initial proving flight to Egypt, but died before the record attempt. This began at Cranwell with a heavily laden 4,500 ft take-off run on 6 February 1933. Bett's place was taken by Flt Lt G.E. Nicholetts, AFC. After 57 hours 25 minutes in the air, they landed at Walvis Bay, South Africa. Their new record of 5,309 miles meant that the UK held all three absolute records for speed, height and distance.

Fairey built the Firefly III, with a supercharged Rolls-Royce F.XIS engine, to take part in a competition for a deck-landing fighter for the Fleet Air Arm. The contract went to the Hawker Nimrod, so the Firefly III was put on floats and used for training by the RAF High-Speed Flight that won the Schneider Trophy outright for Britain in 1931.

Designed as a two-seat naval spotter-reconnaissance/fighter, with an all-steel structure and 520 hp Kestrel IIMS engine, the Fleetwing joined the Firefly III for practice and sea-state/weather checks with the 1931 Schneider Trophy team.

The derivation of the RAF's Gordon two-seat general purpose/day bomber is clear from its initial designation of Fairey IIIF Mk V. The 525 hp Armstrong Siddeley Panther IIA radial engine gave it slightly better performance than the IIIF. Altogether, 246 were built, serving at home and in the Middle East until the outbreak of the Second World War. This one was operated by No. 40 Squadron.

An observer of a 40 Squadron Gordon takes on board a reconnaissance camera. Only the sights of his Lewis gun, stowed on its Fairey high-speed mounting, project above the fuselage. However, the increased width of the radial engine, compared with the IIIF's in-line Lion, meant that the forward-firing Vickers gun had to be mounted externally. Bomb load was 500 lb.

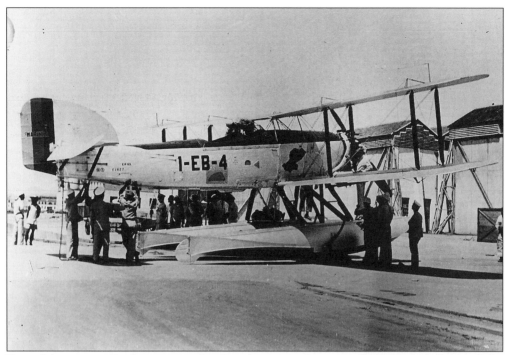

Five of the twenty Gordons exported to Brazil were seaplanes, similar to aircraft flown off the Nile, at Khartoum, by No. 47 Squadron, RAF.

Known originally as the IIIF Mk VI, the Seal served with the Fleet Air Arm as a three-seat carrier-borne spotter-reconnaissance landplane and as a catapult seaplane. The well identified Fairey hangar at the Great West Aerodrome, Harmondsworth, over which this one is flying, retained its company name long after it had been absorbed post-war into the new commercial airport of Heathrow.

A Seal being prepared for take-off at sea. It was the first standard Fleet Air Arm type to have wheel brakes and the first with an A-frame hook for use with newly introduced deck arrester wires. It differed from the Gordon in having a tailwheel instead of a skid.

Lowering a Fairey Seal onto its catapult for launch from a warship of the Royal Navy. Maximum speed in this form was 129 mph and endurance 4.5 hours.

Seals lined up for review by King George V on the occasion of his Silver Jubilee in July 1935. At the time, the Fleet Air Arm had a total of only 175 aircraft in 15 squadrons, many clearly built by Fairey.

Swastikas were unfamiliar markings on British-built aircraft in the 1930s, but were the national symbol of Latvia and Finland. This Seal, with a Bristol Pegasus engine, was the first of four delivered to the Latvian Air Force.

The Fairey G.4/31 was designed to a near-impossible Air Ministry Specification for a multi-role aircraft that would do everything from day and night bombing, daytime dive bombing, army co-operation, reconnaissance, air photography and casualty evacuation to coastal-based torpedo-bombing, if necessary in tropical environments and from unprepared airfields. It was powered by a 750 hp Armstrong Siddeley Tiger engine in its Mk II form shown. The pilot's cockpit was offset to port, making room for a passage from the rear-gunner's cockpit to a roomy cabin housing stations for a navigator and a bomb-aimer, with a side door to admit casualties. Only the prototype was built.

The Kestrel-engined S.9/30 two-seat fleet spotter-reconnaissance biplane first flew on 22 February 1934. Within a few weeks it had to compete with a second Fairey aircraft designed to perform much the same role plus torpedo dropping, the TSR II, which the Air Ministry ordered under the name Swordfish.

The success of the TSR II ended all hope of getting a production contract for the S.9/30. It was converted into a seaplane, with a single 30.5 ft long central float and two underwing stabilising floats like those of the experimental IIIF S1835, and sent to the Marine Aircraft Experimental Establishment, Felixstowe.

First flown by Chris Staniland on 6 June 1935, the Fantôme single-seat fighter is regarded by many people as the most beautiful biplane ever built. Designed to meet a Belgian Air Force requirement, it had two machine guns in the lower wings and two more above the engine which could be exchanged for a 20 mm cannon firing through the propeller hub of the 925 hp Hispano-Suiza 12Yers engine.

The Fantôme's maximum speed of 270 mph far exceeded the Belgian Specification, but it was a biplane in a new monoplane age. Only three more were built, by Avions Fairey, under the name Féroce, of which two ended up fighting with the Republican Air Force in the Spanish Civil War after being exported to Russia.

Unique as the only heavy bomber built by Fairey, the Hendon marked an important milestone as the first cantilever monoplane heavy bomber manufactured in the UK. The prototype was the only monoplane designed to meet Air Ministry Specification B.19/27, which called for a range of 920 miles at 115 mph with 1,500 lb of bombs. It first flew on 25 November 1930, powered by two 525 hp Bristol Jupiter radials.

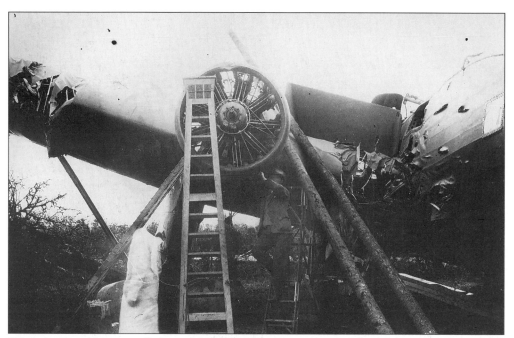

While landing after a test flight on 15 March 1931, the prototype Hendon overshot the airfield, crossed a road and ended in an orchard garden. Damage was so extensive that it had to be virtually rebuilt. The apparently headless man in a white coat moved during the photographer's time exposure.

The rebuilt Hendon had 480 hp Kestrel IIIS engines and the pilot now sat in an open cockpit, with an optional second seat in tandem. By 1933, the Air Ministry was beginning to regret its lack of interest. The Heyford biplane, which it had ordered, was outdated, the new Wellington and Whitley existed only on paper; so it ordered fourteen Hendons.

The production Hendon was by no means ugly, despite its large mainwheel fairings. Two 695 hp Kestrel VIs gave it a maximum speed of 155 mph and it had a range of 1,360 miles. Maximum bomb load was 1,660 lb, and it had defensive gun turrets in the nose, tail and amidships. The pilot's cockpit was again enclosed.

Hendons were flown mainly by No. 38 Squadron, first at Mildenhall and then Marham. One flight was detached as the nucleus of No. 115 Squadron, but orders for 62 more were cancelled and the Hendon was retired before the outbreak of the Second World War.

Three
Arms for Another War

The Hendon signalled the end of the biplane heavy bombers in the Royal Air Force; yet the Fleet Air Arm ordered, at much the same time, an antiquarian biplane that was to win immense glory throughout the Second World War. The Swordfish's victories were superb; its failures, like those of Fairey Battle bombers in 1940, were deeds that sometimes gained Victoria Crosses for their aircrew. Only Swordfish had the slow-flying ability to drop between German battleships racing through the Channel and their destroyer escorts, to launch torpedoes. When Barracudas attacked, later in the war, they dived on their targets and suffered less fearsome losses; but they remained far from beautiful.

Rare for a carrier-based aircraft of its time, the monoplane Firefly two-seat fighter did attain a degree of elegance in its Mk 4 version, but this faded when it went on to the utilitarian three-seat anti-submarine Mk 7. The fighter versions flew most of their combat missions in the Far East and Pacific against Japanese targets in the Second World War, and from carriers of the Royal Navy and Royal Australian Navy during the Korean War. By then, their enemies were flying MiG jets.

The original TSR I of 1933, from which the Swordfish was evolved, was a two-seat torpedo-bomber/three-seat spotter-reconnaissance aircraft intended for the Greek Navy. The engine, at the time of its first flight, was an uncowled 625 hp Armstrong Siddeley Panther VI, soon changed for a 635 hp Bristol Pegasus IIM in a Townend ring.

When the TSR I got into a flat spin and crashed, Fairey built the TSR II, with a longer fuselage, spin recovery strakes forward of the tailplane leading-edge, slightly swept top wings and a 690 hp Pegasus IIIM3. In the spring of 1935, the Air Ministry ordered three pre-series and 86 production aircraft, with the name Swordfish. Eventually, 692 would be built by Fairey and 1,700 by Blackburn Aircraft Company.

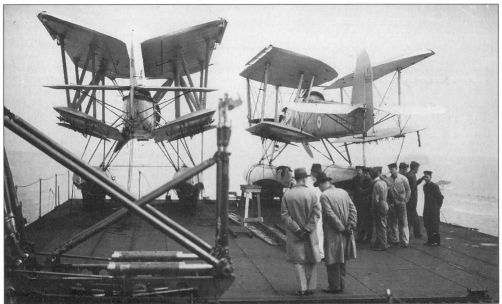

Swordfish (right) and Blackburn Shark side by side on a loading barge during official development trials. The Shark entered first-line service in 1935, but was replaced by Swordfish within three years.

View from the rear cockpit of a Swordfish taking off from HMS *Eagle*. The presence of so many other Swordfish on deck emphasises the short take-off run needed by this antique looking biplane.

By the time of the Coronation Review of the Fleet by King George VI at Spithead, on 20 May 1937, Swordfish had equipped five squadrons of the Fleet Air Arm, generally replacing Blackburn Baffins and Fairey Seals. This formation is flying over the Royal Yacht *Victoria and Albert*.

The most important role of Swordfish units in the final years of the 1930s was training for the war that seemed inevitable. This Swordfish is dropping a practice 18 in torpedo. Alternative weapon loads for the Mk I were a 1,500 lb sea mine, or equivalent load of bombs or depth charges.

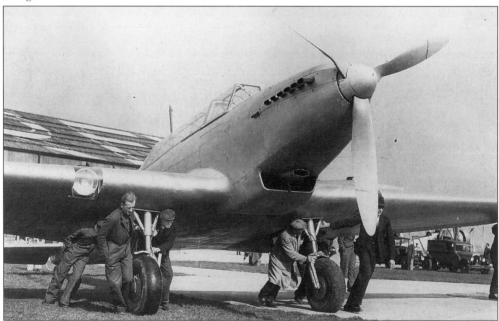

To replace the RAF's Hawker Hart and Hind biplane light bombers, Fairey designed the Battle to Specification P.27/32. Powered by a 1,030 hp Rolls-Royce Merlin engine, it carried a crew of three (pilot, bomb aimer/observer and radio operator/gunner) and 1,000-1,500 lb of bombs. The obvious weakness was that, with the same engine as Hurricane and Spitfire single-seat fighters, its flat-out speed was only 257 mph. Armed with the standard two machine guns, it would clearly be very vulnerable in combat before the end of the 1930s. There was no hint of this when the prototype was rolled out for demonstration to VIPs and the press.

Under Britain's massive rearmament programme of the mid-thirties, the Battle was ordered in large numbers, with Austin Motors' Shadow Factory in Birmingham soon being brought in to supplement deliveries from this Fairey assembly line at its Heaton Chapel, Stockport, plant. Eventually, 2,201 Battles were to be produced.

Battles at Fairey's Great West Aerodrome, Heathrow. The first squadron to fly them was No. 63 at Upwood. By the outbreak of the Second World War there were seventeen RAF Battle squadrons.

Expecting that Fairey's military commitments, under the rearmament programme, would be completed by 1940, the UK Director General of Civil Aviation and Director of Civil Research and Production encouraged Fairey Aviation to venture into the commercial aviation business by awarding it a contract under Specification 15/38. The resulting F.C.1, shown in model form, would have been the most advanced airliner of its day.

Full-scale wooden mock-up of the F.C.1's centre-fuselage and inner wings, in the experimental shop at Hayes. This view shows clearly the Fairey-Youngman auxiliary aerofoil flap under the wing, which would have reduced take-off and landing speeds at the grass aerodromes of the period. It formed an integral part of the wing lower surface when retracted, and contributed later to the performance of the Firefly and Barracuda.

The F.C.1 was to be pressurised, requiring a circular fuselage section. The mock-up emphasised the roominess of the 26-passenger cabin. Normal airline stage length was intended to be 750 miles at not less than 200 mph at 12,000 ft; maximum range was 1,700 miles with reduced payload.

Mock-up of the planned F.C.1 flight deck, complete with Sperry autopilot panel. A flight engineer would have been seated behind the captain on longer flights, responsible for engine, fuel and air-conditioning management.

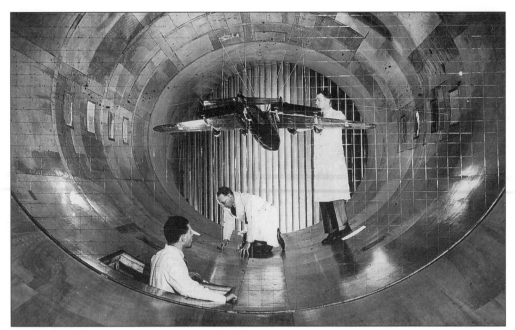

F.C.1 model in the newly built wind tunnel at Hayes, with its Fairey-Youngman flaps extended. Powered by four 1,000 hp Bristol Taurus radial engines, the initial batch of twelve production aircraft would each have cost £80,000, but the contract was cancelled one month after the war started.

Sir Malcolm Campbell, who set world absolute speed records for both cars and boats, with the scale model of his *Bluebird* speedboat that was tested in the Hayes wind tunnel. Such shapes could develop so much lift that they might somersault at high speed, so airflow measurements were vital.

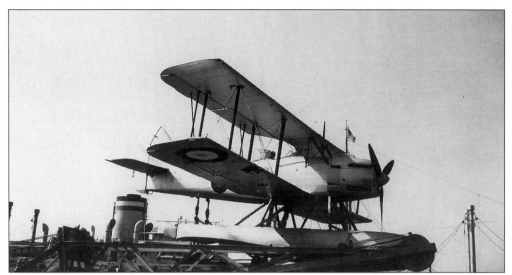

Fairey Seafox two-seat reconnaissance seaplanes, with 395 hp Napier Rapier engines, were carried by both Royal Navy cruisers and armed merchant cruisers during the first four years of the war. One, launched from HMS *Ajax*, spotted for the guns of the three cruisers that defeated the German pocket battleship *Admiral Graf Spee* in the Battle of the River Plate on 13 December 1939. The pilot, Lt E.D.G. Lewin, was awarded the DSC, the first decoration gained by a Fleet Air Arm officer in the Second World War.

Life was proving less satisfactory for crews of the ten RAF squadrons of Battles sent to France with the Advanced Air Striking Force on 2 September 1939. With no bombing by either side during this period of the so-called 'Phoney War', they were used for armed reconnaissance. One rear gunner shot down a Messerschmitt Bf 109 on 20 September, the first RAF combat victory of the war. By 30 September, when all five Battles from No. 150 Squadron were lost on a reconnaissance mission, the aircraft's vulnerability was beyond doubt.

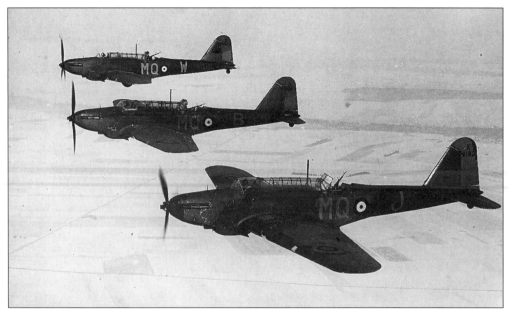

Battles of No. 226 Squadron, one of the units of the Advanced Air Striking Force in France, 1939-40.

Bombing up an AASF Battle. On 12 May 1940, all five aircraft from No. 12 Squadron were lost while trying to stem the German advance through Belgium by attacking bridges over the Albert Canal. The RAF's first Victoria Crosses of the war were awarded posthumously to two members of the aircrew, Flg Off D.E. Garland and Sgt T. Gray. Two days later, 35 of 63 Battles from eight squadrons despatched against enemy troops and pontoon bridges near Sedan failed to return.

After raids from the UK on German invasion fleets gathering in French and Dutch ports, attack missions by Battles ended in October 1940. However, 739 served in Canada and 364 in Australia as dual-control trainers and target tugs, under the Commonwealth Air Training Plan. The trainers had widely separated cockpits for the pupil and instructor.

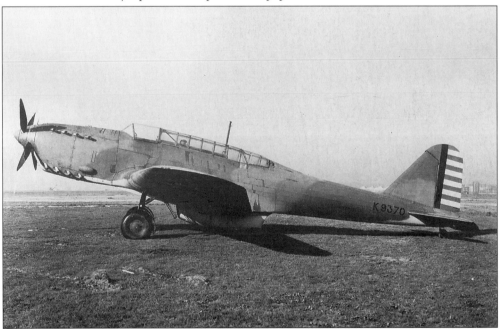

Battles were used as flying testbeds for a variety of new engines. Fairey's own 24-cylinder P.24 Prince 'double engine', consisting of two vertically-opposed banks of six cylinders, each driving one of two contrarotating propellers, was installed in K9370. The US authorities found it so interesting that the aircraft was shipped to Wright Field, Ohio, for a 253 hour evaluation.

King George VI and Queen Elizabeth visited the Hayes works on 15 November 1940. Behind them are directors Maurice Wright and L. Massey Hilton.

Swordfish torpedo-bombers from Malta, never more than 27 in number, sank an average 50,000 tons of enemy shipping every month during one 9 month period. While HMS *Eagle* was in port in Egypt, three of her Swordfish, led by Capt O. Patch, DSO, DSC, Royal Marines, were despatched on 22 August 1940 to attack a submarine spotted in Bomba Bay, Libya, during RAF reconnaissance. Finding two submarines, a depot ship and a destroyer when they arrived, they created such havoc that a planned enemy midget submarine raid on the British Fleet at Alexandria had to be abandoned. The Swordfish in the Fleet Air Arm Museum at Yeovilton is painted to represent Ollie Patch's aircraft.

The most spectacular success came on 11 November 1940, when 21 Swordfish from HMS *Illustrious* attacked the Italian Fleet in Taranto harbour at night. Based on a reconnaissance photograph, this model shows the ships at anchor. After the raid, two of the three severely damaged battleships were underwater, a cruiser and two destroyers hit, two auxiliary ships sunk and seaplane sheds and oil storage tanks destroyed. Two Swordfish were lost.

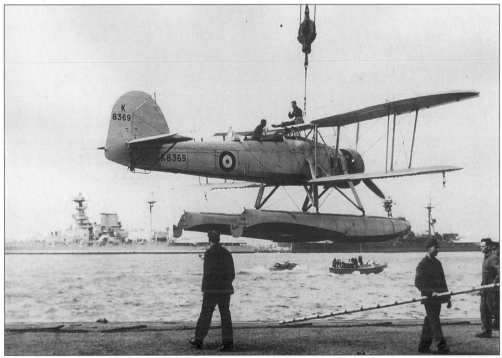

Swordfish seaplanes of No. 701, 702 and 705 Catapult Flights served as spotter-reconnaissance aircraft on board battleships and cruisers. This one is being hoisted ashore in Gibraltar harbour.

Among the most gruelling of Swordfish duties was operation from the carriers that escorted convoys of arms to Russia. After taking off from a snow-covered deck, the crews often had to find their ship in poor visibility on their return. This aircraft is rocket-armed.

Clearing snow from the deck of the escort carrier *Campania* before leaving Murmansk for the return to Britain. On the outward journey, a U-boat had been sunk and attacking German aircraft sent crashing into the sea by the convoy's aircraft and guns, but convoy losses were high.

During a wartime visit to San Francisco, three Swordfish fly over their carrier at anchor below.

The first version of the Swordfish with metal skin on the undersurface of the lower wings to permit rocket launching, was the Mk II, which appeared in 1943. To this, the Mk III (illustrated) added air-to-surface vessel (ASV) Mk X radar. Attacks on U-boats soared. Swordfish from the escort carrier *Vindex* sank four during a single voyage in September 1944.

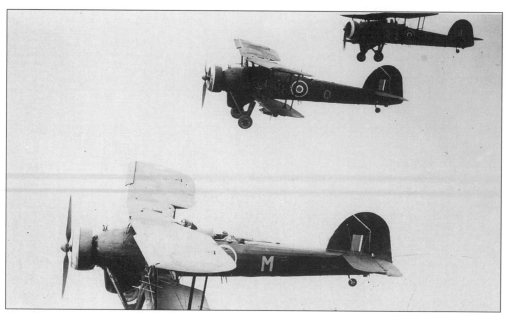

Number 119 Squadron of Coastal Command was the only RAF squadron to fly Swordfish, from October 1944 to May 1945, from Belgian bases. Its final offensive action was an attack on an enemy midget submarine less than four hours before the German surrender.

The Swordfish was intended to be replaced by the Fairey Albacore (left), a much refined product of the same design configuration. In fact, the Swordfish, known affectionately as 'Stringbag' by its crews, outlived the Albacore in operational service.

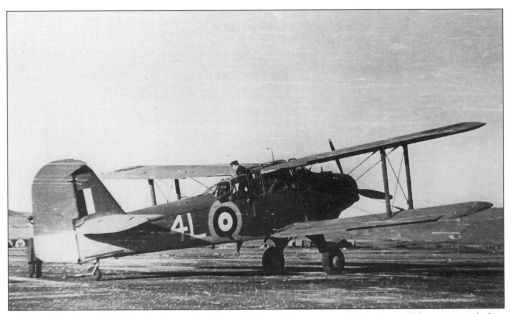

Although the Albacore had enclosed cockpits, the air-conditioning seldom provided a comfortable environment for the three crew inside. First flown on 12 December 1938, it was one of the first British biplanes with an all-metal semi-monocoque fuselage, and 800 were built. Many pilots continued to prefer the Swordfish.

Women workers covering and doping an Albacore wing, with two fuselages being completed to the rear. All Albacores were built at Hayes in 1937-43 and flown at the Great West Aerodrome.

First torpedo attack by Albacores was at the Battle of Cape Matapan in the Mediterranean, in March 1941. Aircraft of No. 826 and 829 Squadrons from HMS *Formidable* battled through intense enemy fire to damage severely the Italian battleship *Vittorio Veneto*. Swordfish of No. 815 Squadron, operating from Crete, helped to disable the cruiser *Pola*.

Albacores were used mainly for land-based operations in the UK and North Africa in 1942-43. Each carrying twenty-eight flares instead of the usual bombs, they illuminated German army units on the ground for the RAF night bombers during the weeks before the crucial Battle of El Alamein in the Western Desert in October 1942.

Symbols painted on the nose of this Albacore fuselage indicate that it sank at least ten enemy ships during anti-shipping patrols. By the autumn of 1943, most Fleet Air Arm squadrons had re-equipped with Fairey Barracudas.

The Fairey P.4/34 prototype was built as a smaller, cleaner and lighter day bomber derivative of the Battle, with a fully-retractable undercarriage. The second prototype (illustrated), which flew on 19 April 1937, was modified in 1938 to become a flying mock-up of a two-seat fighter for the Fleet Air Arm, which went into production as the Fulmar, with eight machine guns and provision for two 250 lb underwing bombs.

A total of 600 Fulmars was built for the Fleet Air Arm. The first 250 were Mk Is, with a 1,080 hp Merlin VIII engine that gave them a maximum speed of 280 mph. The 350 Mk IIs had a 1,300 hp Merlin 30 and were equipped for operation in tropical environments. Operating from HMS *Illustrious* on Malta convoy escort, they started their combat career by shooting down ten Italian bombers in the autumn of 1940 and six more enemy aircraft while providing cover for the Swordfish attack on Taranto. These Mk IIs (foreground) and Grumman Wildcats on the *Illustrious* are accompanying HMS *Valiant* during a practice shoot with her 15 in guns.

During the Second World War, Fairey built Fulmars and Barracudas in its Heaton Chapel, Stockport, works; Beaufighters and Halifaxes were produced at the Fairey-managed Errwood Park Shadow Factory, by the railway line to the left of Heaton Chapel works in this photograph.

Beginning with 25 Bristol Beaufighter Mk IFs, Fairey went on to produce 300 Mk ICs and 175 Mk VICs at Errwood Park.

Of the total 2,602 Barracuda torpedo-bomber, dive-bomber and torpedo-reconnaissance aircraft produced for the Fleet Air Arm, Fairey built 1,192 at its Heaton Chapel works. They comprised two prototypes, 25 basic Mk Is with a 1,300 hp Merlin 30 engine; 675 Mk IIs with 1,640 hp Merlin 32 and ASV Mk IIN radar; 460 Mk IIIs with ASV Mk X radar blister under the fuselage for anti-submarine reconnaissance; and 30 much-changed Mk Vs with 2,020 hp Rolls-Royce Griffon engine, strengthened airframe, taller, pointed rudder and extended, square-tipped wings. All were three-seaters, except for the two-seat Mk V, intended for the Pacific War. Some Mk IIs and IIIs were converted to Mk Vs. Other Barracudas were built by Blackburn, Boulton Paul and Westland.

The Barracuda Mk II was a torpedo-bomber and dive-bomber, carrying a pilot, observer/navigator and radio operator/gunner in tandem. Its radar aerials can be seen above the wings. Armament consisted of two machine guns in the rear cockpit , a 1,610 lb torpedo or 1,500 lb mine under the fuselage, or four 500 lb or six 250 lb bombs under the wings.

The Barracuda first flew on 7 December 1940, but production to replace the Albacore was deferred for nearly two years to concentrate work on desperately needed RAF fighters and bombers. It entered service in January 1943 and made its name on 3 April 1944 with a brilliant attack on the German battleship *Tirpitz* in Kaafjord, Norway. Forty-two Barracudas, escorted by 80 fighters, from 6 RN carriers, achieved 15 direct hits on the warship, starting a great fire. Further Barracuda attacks followed. A wooden ramp was built on the deck of the *Furious*, to make possible take-off with a 1,600 lb bomb. The *Tirpitz* never again went to sea and was sunk eventually with 12,000 lb Tallboy bombs from Lancasters.

A Barracuda II nearing completion at Hayes in August 1944. The aircraft's Fairey-Youngman flaps, seen above the folded wings, had caused such turbulence when lowered on the prototype that the tailplane was raised from the original low-set position on all production Barracudas.

One of the Barracudas that attacked Sumatra on 24 August 1944, when No. 815, 817 and 831 Squadrons from HMS *Indomitable* and *Victorious* heavily damaged an important cement manufacturing plant near Padang, and ships and shore installations at Emmahaven harbour. Positioning the aircraft on deck by pushing on the flaps and tailplane bracing strut would not have been recommended by Fairey designers.

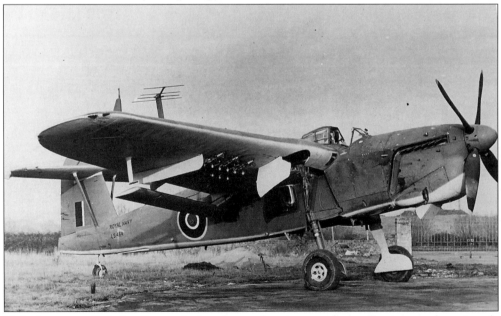

Nobody ever called the Barracuda handsome, but it proved an effective combat aircraft, especially in its dive-bombing role. It dropped torpedoes only rarely.

Barracuda wing folding had to be done manually. With such high-mounted wings, it would not have been possible without the retractable triangular-shape appendage at each wingtip.

Fully-folded and viewed from the rear, the Barracuda looked a nightmare. Only brilliant design could have resulted in a folded 'mighty metal monster' requiring only inches more space than a folded Swordfish.

In flight, without external stores, a Barracuda II had a maximum speed of 235 mph, compared with 139 mph for the Swordfish and 161 mph for the Albacore. This one is flying near Gibraltar.

MX613, a Barracuda II was flight tested with an air-sea rescue lifeboat under its fuselage. Another conducted live drops with paratroops carried in two-person containers under each wing. The containers had side windows, but the idea was judged to be psychologically unacceptable and was abandoned.

When the prototype of Fairey's monoplane Firefly was flown by Chris Staniland from the Great West Aerodrome on 22 December 1941, four wooden 'broomsticks' represented 20 mm guns in the wings. Six months later, the chief test pilot was killed when the tail of the second prototype collapsed at low level, but the Firefly was to prove a major success for its designer, H.E. Chaplin. It remained in production until 1956, in eight major versions. Altogether, 1,702 were built, 132 of them by General Aircraft.

Early Firefly F.1s had a 1,730 hp Rolls-Royce Griffon IIB engine, replaced with a 1,990 hp Griffon XII from the 470th aircraft. They carried a pilot and observer/navigator and were armed with four 20 mm guns, with provision for eight rockets or two 1,000 lb bombs underwing. Maximum speed was 316 mph at 14,000 ft.

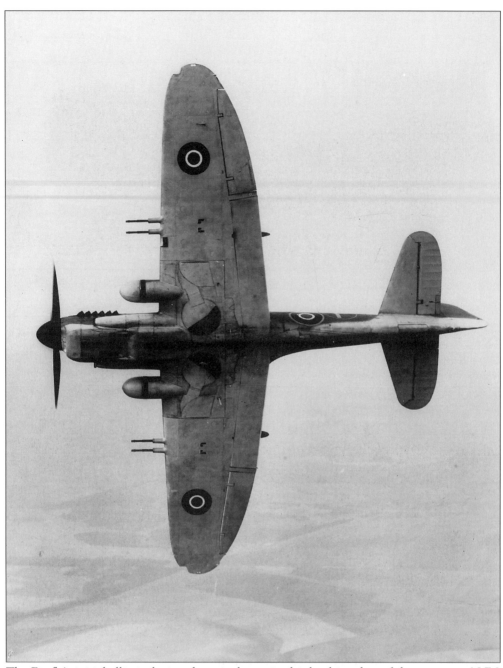

The Firefly's initial elliptical wing shape is shown in this banking shot of the prototype N.F.2 night fighter. Radomes for the ASH anti-ship and anti-submarine radar project ahead of the wing centre-section and the nose had to be lengthened to compensate for the weight of the radar operator's equipment in the rear cockpit. Only thirty-seven N.F.2s were built, following development of more compact radar that could be fitted in a single pod under the engine of the Firefly N.F.1 without any structural modification to the basic F.1 airframe.

Firefly N.F.1, distinguished by the radar pod under its engine and two underwing fuel tanks for extended range.

The first two Firefly squadrons soon made their mark. Number 1770, based on HMS *Indefatigable*, made photographic reconnaissance flights over the battleship *Tirpitz* before its sinking by RAF Lancasters. Number 1771, flying from *Implacable*, attacked enemy shipping off the Norwegian coast. The results are clear from the crews' faces after their return. On the other side of the globe, in January 1945, No. 1770 took part in the great Fleet Air Arm attacks on two oil refineries in Sumatra, curtailing vital supplies to the Japanese. Number 1771, on board *Implacable* as part of the British Pacific Fleet, took part in the attack on Truk in June and became the first Fleet Air Arm unit to fly over the Japanese homeland on 10 July. Between 21 and 30 August, Fireflies were able to drop supplies to prisoner-of-war camps in Japan following that nation's surrender.

Four
The Final Years

Fairey Aviation had experienced serious problems during the Second World War. Albacore production was delayed by difficulties with the Taurus engine; this in turn put Firefly deliveries from Hayes more than a year behind schedule. At Heaton Chapel, Barracuda production slipped by about two years as priority was given to Fulmars, Beaufighters and Halifaxes and its originally planned Rolls-Royce Exe engine had to be changed to a Merlin. At the end of 1942, the newly appointed Minister of Aircraft Production, Sir Stafford Cripps, stepped in. An outside controller was put in to act as deputy chairman and managing director, and Britain's first American-style project team organisation was set up, in which each new Fairey type would have its own design, aerodynamics and structural engineering team. Threatened nationalisation, as had happened at Short Brothers, was avoided.

When the war ended Fairey was in a relatively good position, with continuing orders for Barracudas and Fireflies. The anti-submarine Gannet followed, but the company's bid for leadership in the new technologies of supersonic flight, missiles and helicopter design was insufficient to sustain it in a period when Government policy switched from nationalisation to rationalisation and industry mergers. The Gannet became the last Fairey aircraft to fly on first-line duties with the Royal Navy. The Rotodyne was seen for a time labelled as a Westland helicopter before it died, and with it Britain's last hope of leadership in rotating-wing flight.

Previous pages: A Fairey Gannet A.S.1 from HMS Centaur flies overhead as the carrier refuels at sea from the oiler Tidesurge, off Fraser Island, Queensland, Australia. The escort is the destroyer Lagos.

The Fairey main offices at Hayes in 1950. Beyond them is the research department building, opened in 1938.

Fairey's flight test centre at Ringway Airport, Manchester, 1946. Aircraft on the ground included three Halifax bombers, a Swordfish, a Firefly F.1 and two Barracuda Vs.

The Firefly F.R.4, which entered service with No. 825 Squadron of the Fleet Air Arm in August 1947, was very different from the Mk 1. The new 2,250 hp Griffon 74 engine was enclosed in neat cowling, with the former beard radiator replaced by slim radiators in the extended wing centre-section. Fuel displaced by the radiators went into a large streamlined tank under the port wingtip; the radar went into a similar container under the starboard tip. Maximum speed increased by 70 mph to 386 mph and the rate of roll was increased by clipping the wingtips. Firefly 4s and multi-role 5s, seen here with a single Hawker Sea Fury, played an important part in the Korean War.

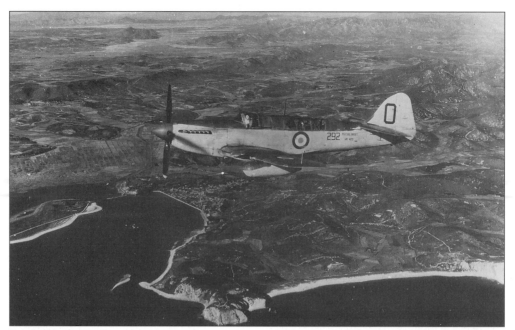

Firefly of No. 825 Squadron from HMS *Ocean* on patrol over the Korean War zone. It was flown by Commissioned Pilot R.B. Wigg, who had logged more than 100 sorties at the time the photograph was taken.

Rocket-armed Fireflies taking off in Korean waters. The 14th Carrier Air Group, comprising 812 Squadron's Fireflies and 804 Squadron's Sea Furies, flew 4,834 operational sorties from the light fleet carrier *Glory*, for the loss of twenty-seven aircraft. Six Firefly squadrons fought in the war from five carriers.

The Firefly 4s of No. 817 Squadron flew from HMAS *Sydney*, the Royal Australian Navy's light fleet carrier. On one mission in support of UN troops they destroyed four key bridges with just sixteen bombs.

Seven Firefly 5s lined up, with the Company Rapide G-AMJK, at White Waltham, which replaced the Great West Aerodrome, Heathrow, as Fairey's main southern flight test centre.

The Firefly A.S.6 was an anti-submarine aircraft without armament, which remained in Fleet Air Arm service until replaced completely by Gannets in 1956. It carried British sonobuoys on its wing and fuselage racks. This one was based at RNAS Ford in Sussex.

H.G. Gregory, General Manager of Fairey's Stockport works, hands over the log book of the first reconditioned Firefly F.1 for the Ethiopian Air Force. Assured that the aircraft would be supplied with full servicing manuals, the Swedish Count Carl Gustaf von Rosen, who re-established the Air Force for the Emperor after the war, asked for them to be delivered with nothing but national markings and serial numbers painted on the camouflage. The Fireflies looked immaculate when flown from Ringway to Ethiopia by pilots trained under his supervision.

Svensk Flygtjanst of Sweden was contracted to tow glider and sleeve targets for practice by Swedish anti-aircraft batteries. It suggested that surplus Firefly 1s would be ideal for the work if fitted with an RFD windmill-operated winch, and this was the first of a dozen or more Firefly T.T.1 conversions that it bought. Others went to Denmark and the Indian Navy.

More than 100 Firefly 1s were converted into pilot trainers from 1946 by installing a dual-control rear cockpit for an instructor, raised to ensure a good forward view. This was one of twelve T.5s, similarly converted into trainers by Fairey Aviation of Australia for the Royal Australian Navy.

The Firefly 7 was the final version produced by Fairey, except for thirty-four Firefly U.8 pilotless targets. The basic A.S.7 illustrated was intended to meet Fleet Air Arm requirements for an anti-submarine search aircraft until the Gannet was ready for service. It reverted to a large beard radiator and full-span wings, had a taller fin and a blister-canopy rear cockpit for two radar operators. No armament was carried, but a flat container for sonobuoys was attached under each wing. The A.S.7 saw only limited service.

The Fairey Aviation works at Hayes in 1946, photographed by the Air Survey Company, one of its subsidiaries.

The winter of 1946/47 was extremely severe, with heavy snowfalls. Cuts in electrical power supplies were frequent and unpredictable. Fairey brought in tractors to keep its Hayes machine shop at work.

Founded at the Stockport works in 1937, the Fairey Aviation Works Band, conducted by the great Harry Mortimer, began its highly successful post-war career by becoming the first winner of the *Daily Herald* National Brass Band Championships at the Royal Albert Hall, London, in 1945. Although Fairey Aviation has gone today, the band continues to prosper as the Williams Fairey Engineering Band.

One of the largest single-engined aeroplanes of its day, the Spearfish was designed under H.E. Chaplin as a two-seat torpedo/dive-bomber with an internal weapons-bay big enough to accommodate an 18 in or 22.4 in torpedo, one 1,600/2,000 lb or four 500 lb bombs, four depth charges, thirty large flares or a 180 gallon auxiliary fuel tank. The first of four prototypes was flown by F.H. Dixon on 5 July 1945. The war ended the following month and orders for 152 production Spearfish were cancelled.

Don't look now, but... as a Spearfish prototype takes off, its 2,000 hp Bristol Centaurus piston-engine roaring at full throttle, a diminutive Vampire fighter, its camouflage making it almost invisible, symbolises the beginning of the jet era that has made the Spearfish already obsolete.

The first product of Fairey's newly formed Research and Armament Development Division (later Weapon Division) at Heston was the Stooge guided missile, conceived in 1945 as a counter to Japanese suicide aircraft. Only 82 inches in span, it was a glide missile, boosted from a ramp to a speed of over 500 mph by four 75 lb jettisonable rockets. This shows one of the successful tests at the Rocket Experimental Establishment, Aberporth, in March 1947. No production order was received.

The Fairey Fireflash was the first air-to-air missile to serve with the Royal Air Force. Propelled by two booster rockets, it was a beam-rider, guided by the radar of its launch aircraft, and was used as a training weapon on the Supermarine Swift F.7s of No. 1 Guided Weapon Development Squadron at RAF Valley in 1957-58.

The prototype of the Fairey Gannet anti-submarine hunter-killer was built at Hayes and had to be towed behind a motor tractor along public roads to Aldermaston, near Reading, for reassembly and flight. Even with the span reduced to 19 ft 6 in by the folded wings, this presented problems.

The prototype Gannet was known initially as the Q or Fairey 17, having been developed to Specification G.R.17/45. It had two seats and was powered by a 2,950 ehp Armstrong Siddeley Double Mamba, comprising two Mamba turboprops coupled through a gearbox to drive coaxial contra-rotating propellers. One engine and one propeller could be stopped in flight to reduce fuel consumption and extend range. The first flight, by Fairey's chief test pilot, Grp Capt Gordon Slade, was at Aldermaston on 19 September 1949.

When the prototype Gannet was guided to a safe touchdown on the deck of HMS *Illustrious* at sea, on 19 June 1950, it was the first time that a turboprop aircraft had landed on board a carrier.

The Gannet was ordered into production with a third cockpit behind the wings, an extended weapons-bay and two auxiliary fins on the tailplane. The fuselages were built on Fairey's unique new envelope jigs, in which the large skin panels were made up inside precisely contoured jigs to ensure their correct aerodynamic form.

Sir Richard Fairey hands over log books for the first Royal Navy Gannet unit, No. 703X Flight, to Lt Cdr F.E. Cowtan, its commanding officer, at RNAS Ford on 5 April 1954. Guests in the background are Fairey Directors G.W. Hall and Richard Fairey, and Insp V.C. Taylor.

While the author waits in the centre cockpit, test pilot David Masters climbs into WN370, the first Stockport-built Gannet, for a spirited display before local dignitaries and the press on 5 October 1954.

WN354, one of the first production batch of 26 Gannets, over the Needles lighthouse, Isle of Wight, with its ASV surface search radome extended. Excluding prototypes, 349 Gannets were built, 78 of them at Stockport, the rest at Hayes.

His years of active service as a Naval officer gave the Duke of Edinburgh special interest in the Gannet when he inspected the production line at Hayes in October 1954.

Number 816 and 817 Squadrons of the Royal Australian Navy were equipped with Gannet A.S.1s at RNAS Culdrose, Cornwall, in August 1955, before embarking in HMAS *Melbourne* for the passage to Australia.

Fifteen Gannet A.S.4s, with 3,035 ehp Double Mamba 101s, were delivered to the West German Naval Air Arm in 1958. This uprated version could carry two homing torpedoes and three depth charges, or two mines and three depth charges, or one 2,000 lb bomb, two 1,000 lb bombs or four 500 lb bombs, together with sonobuoys, markers and flares. Alternative armament of 16 to 24 rockets could be carried externally under the wings.

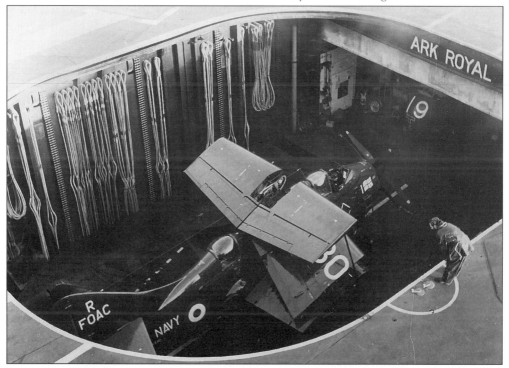

Finished in RAF blue-grey gloss overall, this Gannet C.O.D.4 was used by Flag Officer Aircraft Carriers (FOAC on fin) in 1963. On the wall beside *Ark Royal's* deck lift are strops by which aircraft were attached to the ship's steam catapults for take-off.

Ernest Tips, the Managing Director of Avions Fairey in Belgium, designed and built a series of light aircraft at Gosselies during the second half of the 1930s. They were not Fairey types but a Tipsy Trainer, similar to this Tipsy B prototype, with a 62 hp Walter Mikron engine, was flown post-war by Fairey employees.

Tips called his single-seat Nipper a '1950s Blériot monoplane'. The prototype flew on 3 December 1957, and a few more were built at Gosselies before manufacturing rights were sold to another company. With a 55 hp modified Volkswagen engine, a Nipper could cruise at 93 mph and was aerobatic.

This Tipsy Junior single-seat ultralight was imported into the UK by Fairey Aviation in 1948 to investigate whether or not a market existed for an aircraft that spanned less than 23 ft, had an empty weight of 486 lb and was powered at that time with a 36 hp J.A.P. Before flying it, Hawker test pilot Sqn Ldr Frank Murphy, DFC, received a briefing from F.H. Dixon, who had been Fairey's chief test pilot from 1942 to 1945 before assisting with development of the Gyrodyne helicopter.

Parts of the Tipsy M, built at Gosselies in 1938-39 with a Gipsy Major engine, were used by Fairey Aviation to produce this two-seat Primer trainer in 1948. A second Primer, with a 155 hp Blackburn Cirrus Major, was built to compete against the de Havilland Chipmunk for an RAF trainer contract. The Chipmunk won and no more Primers were made.

In August 1945 Fairey entered the rotary-wing business by forming a team to build a revolutionary aircraft known as the Gyrodyne, proposed by Dr J.A.J. Bennett. Members of the team, shown here with the prototype, are from left to right, standing: Ken Poulton, R.D. Trumper, L. Baker, G.B.L. Ellis, Dr J.A.J. Bennett, E. Parsons, J. Oliver, J. Dempsey and F.L. Hodges. Seated: Derek Garroway, Basil Arkell, F.H. Dixon, S. Verge and J. Gillan.

The Gyrodyne had a three-blade rotor driven by a 520 hp Alvis Leonides engine. During take-off, landing and hover, a propeller at the starboard wingtip absorbed only sufficient engine power for yaw control. In cruising flight, most of the power went to the propeller for propulsion, leaving the lightly loaded rotor to provide only lift, within or near autorotative pitch. The prototype, G-AIKF, piloted by Basil Arkell, set a world helicopter speed record of 124.3 mph over a 3 km course at White Waltham on 28 June 1948. On 17 April 1949, F.H. Dixon and Derek Garroway were killed when the rotor head suffered fatigue failure during practice for an attempt on the 100 km closed-circuit record. The second prototype was grounded, to reappear four years later as the much changed Jet Gyrodyne.

In the Jet Gyrodyne the two-blade rotor was turned during take-off, landing and slow-speed flight by fuel-burning pressure-jets at the tips. Air for the pressure-jets was fed through the blades from two Rolls-Royce Merlin compressors driven by the Leonides engine. In cruising flight, the rotor was unpowered and the Leonides drove two pusher propellers at the wingtips. More than 190 transitions from helicopter to autogyro flight and 140 autorotative landings were made during flight testing.

Pressure-jets, supplied with air by a Palouste turbojet, powered the 28 ft rotor of the two-seat Ultra-light Helicopter, first flown at White Waltham by Sqn Ldr W.R. Gellatly on 14 August 1955. Six were built to meet an Army requirement for reconnaissance, casualty evacuation and training. They landed on small platforms on ships pitching and rolling in heavy seas, carried stretchers during Army exercises and much more - but development ended in 1959.

Forty tests of 10 ft span rocket-powered models were made by Fairey as the first stage of a programme to develop a vertical take-off fighter for the Royal Navy. Each chamber of the Beta 1 liquid-fuel rocket engine developed 900 lb of thrust, and the models were boosted from their launch-ramps by two solid rockets.

One of the Fairey VTO models leaving its launcher at Woomera rocket range, Australia. Earlier tests were made at Aberporth in Wales and from a tank-landing craft at sea, where handling of the rocket fuels was considered too hazardous.

The diminutive Fairey Delta 1 (F.D.1) single-seat research delta was built on the basis of the VTO model tests. It spanned only 19 ft 6½ in, was powered by a Rolls-Royce Derwent turbojet and had a take-off weight of 6,800 lb.

The F.D.1 was first flown by Grp Capt Gordon Slade at Boscombe Down on 12 March 1951. Intended originally for ramp launching and as a fighter to counter attacks on ships by suicide aircraft, it had large vertical tail surfaces and elevons for control at low speed after launch. Together with the small wings, this made it difficult to fly in anything but calm conditions, but the rate of roll demonstrated by test pilot Peter Twiss was almost 500 degrees a second.

The small T tailplane of the F.D.1 was fitted only to ensure adequate stability during early flight tests and was to be removed later. It restricted maximum speed to 345 mph, instead of the designed 628 mph, but the little Delta never had a chance to show its true capability. Its undercarriage collapsed after the aircraft swung during landing by a service pilot, and the programme was terminated.

The Fairey Aviation Board of Directors at the Hind's Head Hotel, Bray, on the 40th anniversary of the company in 1955. Seated from left to right: Sir Richard Fairey, Wilfred Broadbent and R.T. Outen. Standing: Maurice Wright, G.W. Hall, L. Massey Hilton, Richard Fairey and C.H. Chichester Smith.

The Delta Two was the last fixed-wing aeroplane designed and built by Fairey Aviation. It had the smallest practical airframe that could be built around a Rolls-Royce Avon RA 28 afterburning turbojet, a pilot and enough fuel for a range of 830 miles; wing span was only 26 ft 10 in. WG774, the first of two Delta Twos, was first flown by Peter Twiss at Boscombe Down on 6 October 1954. On 10 March 1956 he used it to set a world speed record of 1,132 mph. This exceeded the previous record by 38 per cent, a margin that has never been bettered.

After setting his speed record, Peter Twiss is congratulated by R.L. (now Sir Robert) Lickley, Fairey's Chief Engineer, and Maurice Childs, Chief Flight Development Engineer.

Peter Twiss with a special award that he received after his record flights. Because he had to leave his hotel before breakfast was served on the morning of the attempt, the night porter boiled him an egg in a kettle. The press commented afterwards that he had 'set the record on an egg', so the ground crew all signed this egg and presented it to him, suitably mounted by Norman Parker, during a celebration party in the Wheatsheaf at Woodford, near Salisbury.

Delta-wing aircraft normally land nose-high. To ensure a good forward view for pilots of the Delta Two at that time, Chief Designer H.E. Chaplin proposed that the nose, complete with cockpit, should be able to droop 10 degrees hydraulically, a solution copied later on Concorde. WG774, with the record figure painted on its side, lands with nose drooped at the 1958 SBAC Display, Farnborough.

Fairey received no contract for another fixed-wing aircraft as a result of WG774's achievement, but the aircraft contributed later to the superb flying qualities of Concorde when British Aerospace converted it into the BAC 221. This had small scale versions of the airliner's sharp-nosed ogival wings to confirm theoretical and wind tunnel predictions.

Members of the Fairey Long Service Association at a reunion at Hayes on 14 March 1959. Working for the company proved a job for life for many people. Ten years earlier, 282 employees with more than 25 years of service had attended a dinner in London, including 35 who had been with Fairey for at least 30 years.

Fairey Marine, founded at Hamble in 1946, quickly became a national leader in the design and manufacture of small sailing and racing craft. Queen Elizabeth the Queen Mother and Princess Margaret showed particular interest in the company's hot-moulded hull construction during a visit to its stand at the London Boat Show.

In September 1956, the Rotodyne helicopter seemed to offer the brightest hope for the company's future. The prototype, under construction at Hayes, was powered by two 3,000 ehp Napier Eland turboprops. It was designed to take off as a helicopter, its rotor turned by pressure-jets at the blade tips; these were supplied with air by compressors driven by the Elands. At a safe height the Rotodyne was intended to convert into an autogyro, with the rotor autorotating, the wings providing much of the lift and the turboprops driving propellers for forward propulsion.

Before the Rotodyne was cleared for its first flight on 6 November 1957, it completed 50 hours of rotor testing and 100 hours of engine running in this test rig at White Waltham.

Sqn Ldr W.R. Gellatly and Lt Cdr J.G.P. Morton flew the Rotodyne throughout its four years of development and demonstration. On 5 January 1959, they set a speed record of 190.9 mph around a 100 km closed circuit in the FAI's new E.2 class for convertiplanes, exceeding the equivalent helicopter record by 49 mph.

The prototype Rotodyne was flown into London and Paris heliports. New York Airways and British European Airways both showed interest in a larger production version with 5,250 shp Rolls-Royce Tyne turboprops and 54/65 seats.

The cabin of the Rotodyne was large enough to accommodate such vehicles as Land Rovers, driven up a ramp through its rear clamshell doors.

The pressure-jets made the Rotodyne noisy. It was argued that, even if it proved unacceptable for services into city centre heliports, noise was less of a factor if it were put into military service or used as a civilian flying crane, hauling awkward loads in open country. But after Fairey's helicopter activities were taken over by Westland Aircraft in 1960, the Rotodyne programme was abandoned within two years.

Forty-four Gannet A.E.W.3s of No. 849 Squadron were the last Fairey aircraft to serve with the Royal Navy. The radar under the redesigned fuselage was intended to provide early warning of enemy aircraft approaching the Fleet. Sadly, Gannet 3s had been retired by the time they might have saved Royal Navy ships from destruction by Argentine aircraft during the Falklands Campaign in 1982.

When the new HMS *Ark Royal* sailed into Portsmouth in 1985, she was visited briefly by this surviving example of a great Fleet Air Arm aircraft that had served in the *Ark Royal* of the Second World War. Swordfish III NF389 seems to be turning in salute to an even older veteran, HMS *Victory*, as it is lowered on deck.

Side by side on the stern of HMS *Ark Royal* are the Swordfish symbolising Fleet Air Arm gallantry in 1939-45, and the V/STOL BAe Sea Harrier that had dominated the skies over the Falkland Islands in 1982.